How Lucky We Are

Written by
James Wilson

Illustrated by
Yorris Handoko

Illustrated by Yorris Handoko

Hardcover ISBN: 978-1-959202-09-7
Paperback ISBN: 978-1-959202-08-0

Library of Congress Control Number: 2023902944

Lella Menged LLC
5900 Balcones Drive
Suite 100
Austin, TX 78731

2222 W. Grand River Ave,
STE A,
Okemos, MI 48864

www.lella-menged.com

Dedication:

This book was written especially for my fourteen beautiful Grandchildren.

To my wife, Christina, for her unconditional support to myself and family.

Her support has been the foundation for all that I have accomplished in life.

To my father. Arthur Wilson, for teaching me about physics and the cosmos. My first look through my father's telescope at a very early age was awe inspiring. He also taught me to question how everything works and introduced me to flying.

To my mother, Doris Wilson, who was my moral compass and spiritual guide, for teaching me to love and appreciate animals and nature.

To my oldest son Adam, for his editorial inputs, creative thinking and support.

Acknowledgement:

I want to thank my friend Alem Aweke for inspiring me to author this book, and his masterful talents as my publisher. Additionally, much credit is given, to Yorris Handoko for his amazing illustrations and other contributions.

Chapter 1:
Everything Is Just Right

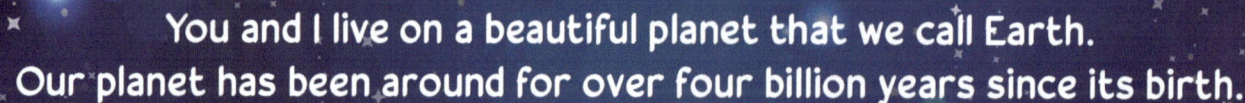

You and I live on a beautiful planet that we call Earth.
Our planet has been around for over four billion years since its birth.

Most of that time, it was too hot, with volcanos and no air.
You see, our Earth is like no other planet and is very rare.

For us to be here, billions of things had to happen, both near and far.
Earth is truly a miracle, so let's explore just how lucky we are!

Our extra-large moon is there for many good reasons.
Our moon stabilizes the Earth's spin so we can have our seasons.

The moon pulls on oceans
to create the tides in the sea.
That helped life to form with
so many creatures to be.

Long time ago, a huge object crashed onto the Earth.

The debris circled the Earth and became...

...the Moon we know and love today.

Our moon once was a planet that hit the Earth and created our spin.

How lucky we are: just the right spin to make day and night again.

Our Sun is so powerful,
with rays that could harm us all,
But our spinning iron core
creates a magnetic shield that stands tall.

This shield protects our planet
and all life, including you and me.
Planets without a spinning
iron core couldn't have life,
not even a tree.

EARTH
MAGNETIC
SHIELD

Chapter 2:
Nebula Flavored Ice Cream

There are many tiny pieces that come together to make you and me.
All of those tiny pieces come together just right so we can live and be.

But where do all these tiny pieces come from that make us who we are?
You might be surprised to learn that our pieces came from a star.

Yes, without the stars, there would be no pieces to make us or anything at all.
No trees or bees, no cats or bats, no plants or pants, not even a ball.

There would be nothing both near and far.
Thank goodness for stars - how lucky we are.

I heard these tiny pieces are called "atoms".

And is our ice cream also made of them?

Yes, and yes.

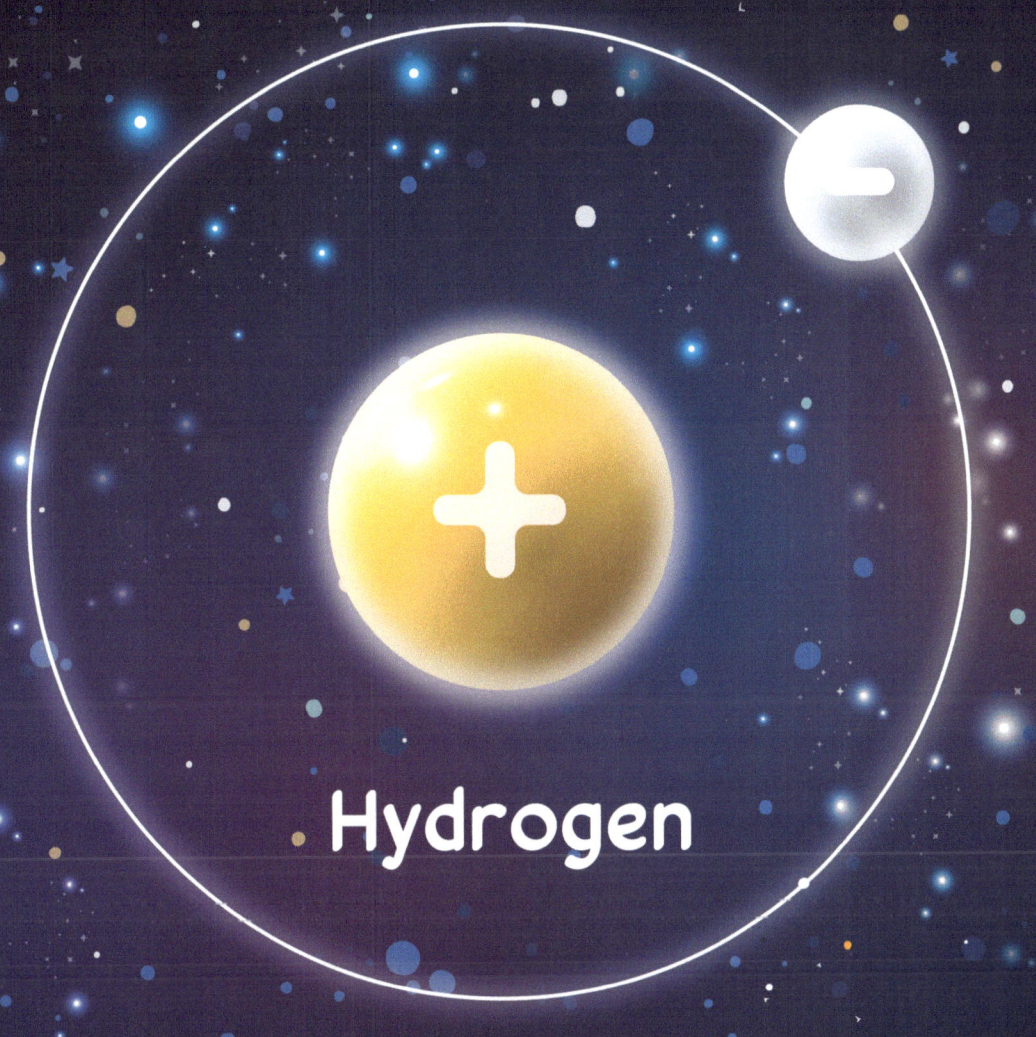

Hydrogen

When the universe first started, it made but just one part.
It was a simple atom called hydrogen. It was a good way to start.

The hydrogen then pulled together to make stars all over the place.
So in the beginning, there were billions of stars and mostly empty space.

A Nebula condenses
into a star.

The new star
burns bright.

Near the end of its life,
the star expands into
a red supergiant.

Other new pieces came together,
and some new stars came to be.
The new stars burned a very long time
and then also exploded, you see.

And more new pieces were made
to create so many other things.
All these pieces make the Earth,
you, me, and even a bird that sings.

The star explodes
and spreads elements
far and wide.

Stars made everything, both near and far.
Thank goodness for them – how lucky we are.

So on the next clear night, look up to the brightest star.
And just say thank you and think how lucky you are!

Chapter 3:
Stardust Superstars

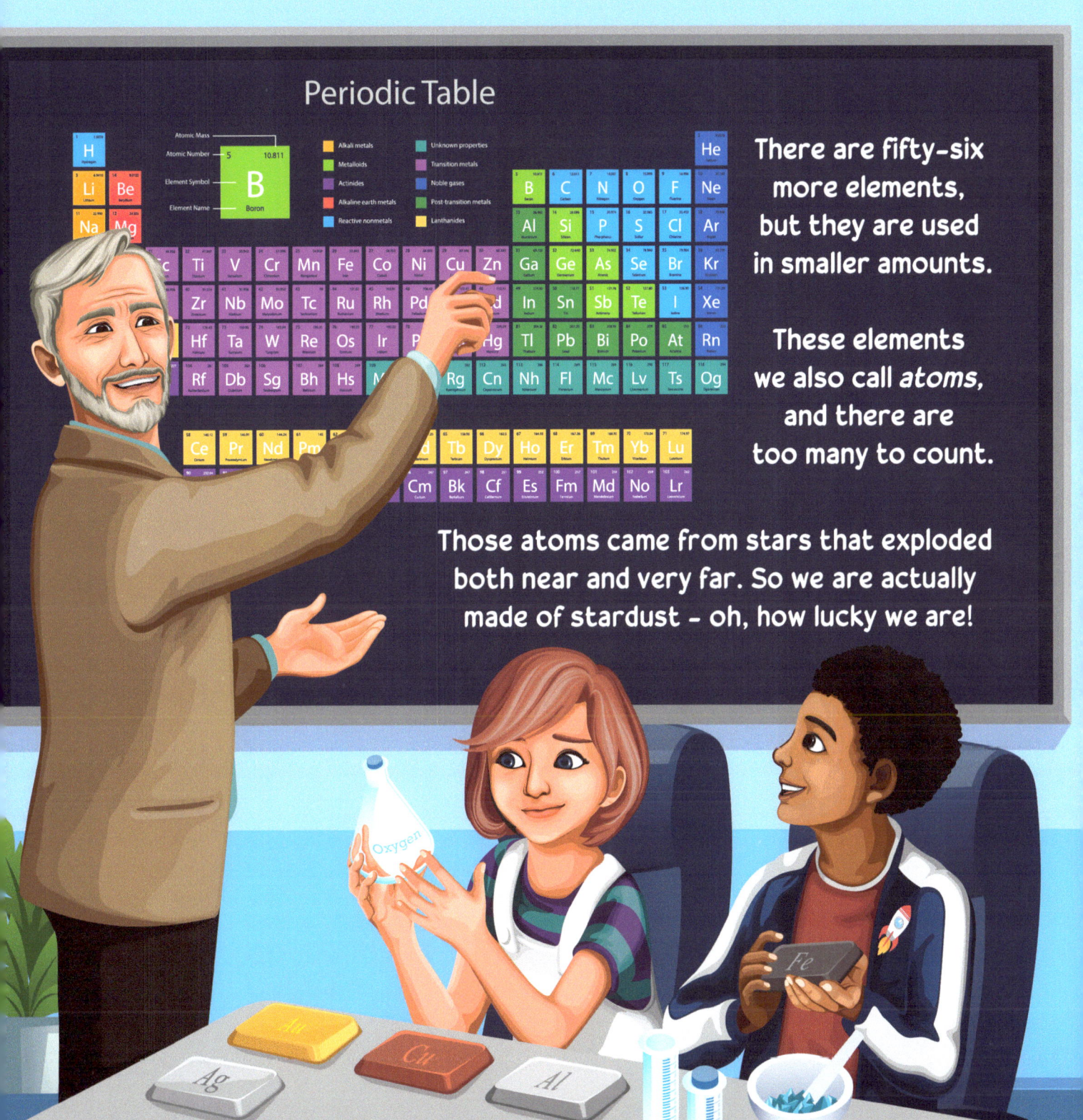

We eat food that our bodies break down into elements for us.
We should eat healthy foods, and that's why your parents make such a fuss.

Our bodies move just the right elements to where they should go:
Carbon for your organs and calcium for the nail that's on your toe.

So eat your vegetables and fruits to take care of your greatest prize.
You're the most fantastic thing in the whole universe, no matter your size.

Chapter 4:
The Greatest Gift

We have learned a lot about
how lucky we are that we came to be.
So many things had to happen –
it's truly a miracle, you see.

Because space is really
a dangerous place with
all its harmful rays,
But our Sun's solar winds
shield our entire solar system
in so many ways.

Gravity and all the other forces in the universe
are balanced in every way.
They keep everything in perfect order
for us as we work and play.

We could go on and on about
so many other things that
had to be just right.
Someday, look through a telescope
and see your universe,
such a wondrous sight.

Our universe is so huge,
we can't even begin to imagine its size.
Billions of things had to happen in perfect order
to make you, the greatest prize.

So when things aren't going so well
and you're making quite a fuss,
Just think about the universe and
how all the big problems have been solved for us.

Enjoy planet Earth, your beautiful home,
a true miracle that came to be.
Think of all that has been created for you.
You are so lucky, you see.

Some people get sad because they don't have certain things,
like big fancy cars. But we all have the whole universe
and everything in it, including the stars.

Too many people think and worry about
what they don't have and always want more.
But we all have the greatest gifts:
the universe, the Earth, and its spinning iron core.

The universe is yours, and you are part of it in so many ways.
I hope you realize how really lucky you are in all of your days.

So now that you know that the awesome universe
is the greatest gift for you and me,you should
appreciate everything in it, and be the very best that you can be.
You should eat healthy foods, learn something useful,
and make yourself proud to be. Then do something
to make our beautiful planet even better on land or in the sea.

You are the most awesome collection of stardust,
with the ability to think and do.
You can make your world an even better place,
it's really all up to you.

www.ingramcontent.com/pod-product-compliance
Lightning Source LLC
Chambersburg PA
CBHW041604120626

46551CB00002B/300